THE
THEORY OF
INTERPLANETARY
EVOLUTION

GERALD S. NORDÉ, JR.

Published in the United States of America by
Do The Write Thing of DC
56 T Street, NW
Washington, DC 20001
www.dothewritethinginc.org

ISBN: 978-1500519230

Book & cover design by Gloria Marconi Illustration & Design

Quilting The Black-Eyed Pea (We're Going to Mars)
©2002 by Nikki Giovanni

ABOUT THE AUTHOR

Gerald S. Nordé, Jr. is a poet, writer, educator and science fiction enthusiast. He is the Founder of Do The WriteThing Foundation of DC (DTWT) which provides poetry writing, songwriting, publishing and video production workshops for youth. Under his guidance, DTWT youth have published four books of poetry. The last two books, *Our Planets, Ourselves* and *Poetry In Motion: The Birth of Success* are their poetic response to his Theory of Interplanetary Evolution.

He earned his Bachelor's degree at Central State University in Wilberforce, Ohio and his Master's degree at Mercy College in New York.

INTRODUCTION

The Theory of Interplanetary Evolution (TIE) is a new theory which offers an innovative perspective on how life evolved by making a connection between human beings and the planets in our solar system. TIE is the theory that all planets in the solar system are interconnected by undergoing the same process of change and growth as humans do, through evolution. TIE uses a "solar birth" model in which our star, the Sun, gives birth to an unnamed planet, very near to its own surface through a tremendous solar flare. That planet begins life orbiting very close to the Sun. During its multi-billion year lifetime, that planet's orbit expands, moving the planet away from the Sun.

As the planet moves further and further away it goes through geological and atmospheric changes or physical phases similar to the phases of life that a human being experiences. The 10 phases or stages the planet experiences are as follows: (1) Mercury, (2) Venus, (3) Earth, (4) Mars, (5) Asteroid Belt, (6) Jupiter, (7) Saturn, (8) Uranus, (9) Neptune and (10) Pluto. Each of these phases corresponds to a 10-year phase in the life of a human being. Mercury represents birth-10 years old; Venus represents 10-20 years old, and so on to Pluto which corresponds with 90-100 years old and death. I have developed a unique form of free verse called the solar cycle poem which uses the structure of the solar system as a metaphor for the stages of human life. See my solar cycle poem, Foreverland, on page 3.

TIE takes into consideration many variations and questions about the existence of life and also proposes the next stage of human evolution. If human beings are on a planet during the Earth phase, as it transitions to the next phase, Mars, then perhaps humans are evolving into Martians.

—*Gerald S. Nordé, Jr.*

*"I think of space
not as the final frontier
but as the next frontier.
Not as something
to be conquered
but to be explored."*

—Neil deGrasse Tyson
(Black American Astrophysicist)

FOREVERLAND
A Solar Cycle Poem
Gerald S. Nordé, Jr.

What's done in the dark, comes to light
Your brilliant star births life sight

Your existence begins and things move fast
From embryo to toddler, experience impacts with a blast

Sometimes it's hard to see what your elders are trying to do
But love is the plan as your young heart breaks through

You come into your own mind, with ego bold and confident
Let the mystery begin, are you truly heaven sent?

You search, you scramble, anxious to discover an explanation
And just when you have it all figured out, you suffer
        a huge devastation

Everything you knew, everything you learned, is tested to no end
And if you make it through, as you've done before,
        eternity will be yours to win

Glory, Fame, Strength, The Universe is yours for the taking
God is you, the opposite is true and your true destiny
        is now awakened

You can hear the soul of spirits past and filter out the madness
You understand the heartbreak, joy, crazy love, even sadness

No more drama, no more baggage, you expel all taunts and sass
Even as a God, you are mystified by what will soon come to pass

You arrived; you conquered the biggest dreams of yours and others
Cosmically insane you are choking, celestial dust, smothered

Your eternal soul is still and rises above the fray
Feast your eyes on the vast foreverland where your dreams
        went to play

What more can be said or done, one is all, all is one
Countdown, Rewind, Begin Again at the Sun

LET'S LOOK AT SOME PROVEN FACTS ABOUT THE EARTH AND THE SOLAR SYSTEM

In 1969 the first men landed on the Moon; Buzz Aldrin and Neil Armstrong. During the next 3-1/2 years, 10 other astronauts would set foot on the Moon's surface.

While the astronauts conducted many experiments, they began one in particular, "Lunar Laser Ranging", which is the key component of a new theory, TIE *(Theory of Interplanetary Evolution)* explaining the next step in the evolution of humankind and the Solar System.

The astronauts placed reflective panels called **"retroreflectors"** on the Moon's surface.

Lasers on Earth were aimed at them and scientists measured the time it took the laser light to reflect back. This would indicate the distance between Earth and the Moon. Over several years they found it was taking longer and longer for the light to return.

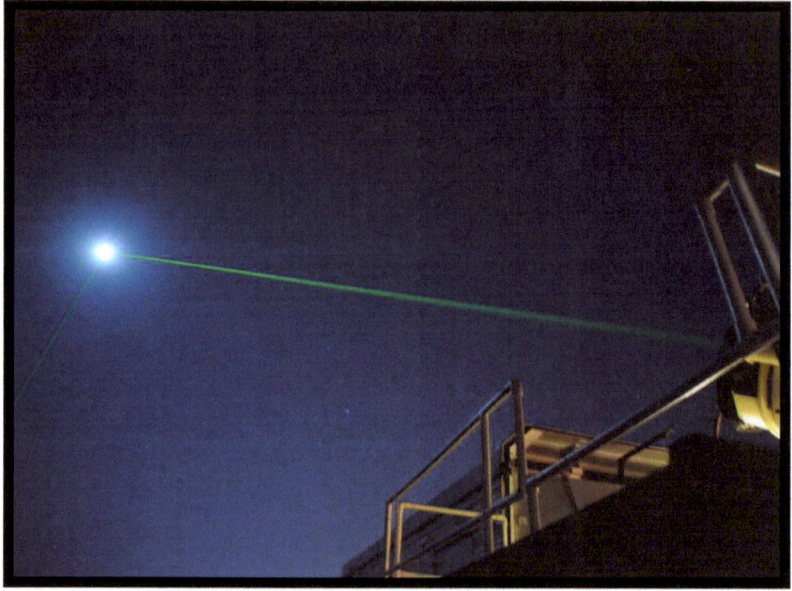

THE LIGHT IS TAKING LONGER
AND LONGER TO REFLECT
BACK TO EARTH FROM
THE MOON.

What does that mean?

RESULTS

This long-term experiment found that the Moon is spiraling away from the Earth at a rate of 3.8 cm per year. This rate has been described as **ANOMALOUSLY HIGH!!!!!**

COULD THE ORBIT OF PLANETS BE EXPANDING, MOVING THEM FURTHER AWAY FROM THE SUN?!

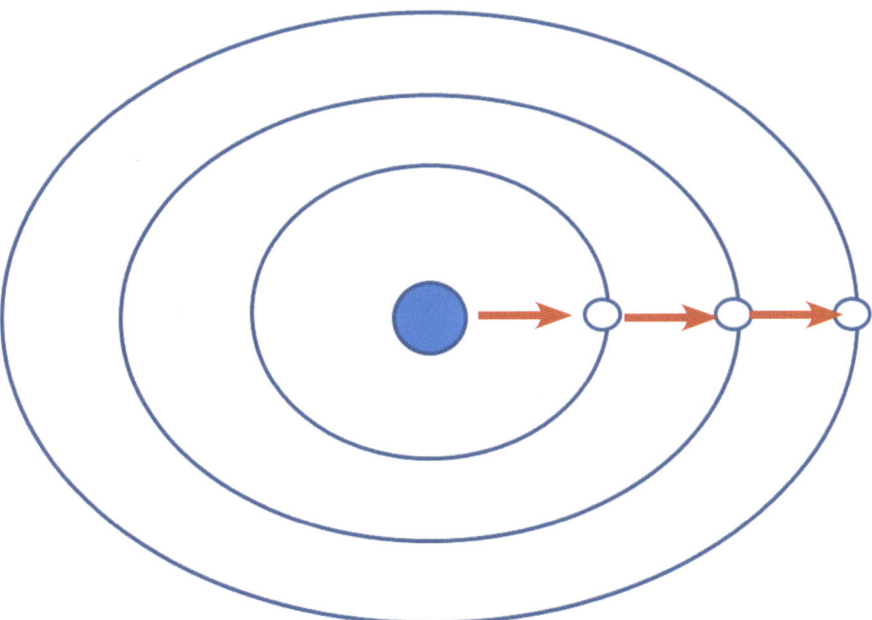

Graphic model displays planet orbiting the Sun. As its orbit expands, the planet moves away from the Sun while undergoing evolutionary changes.

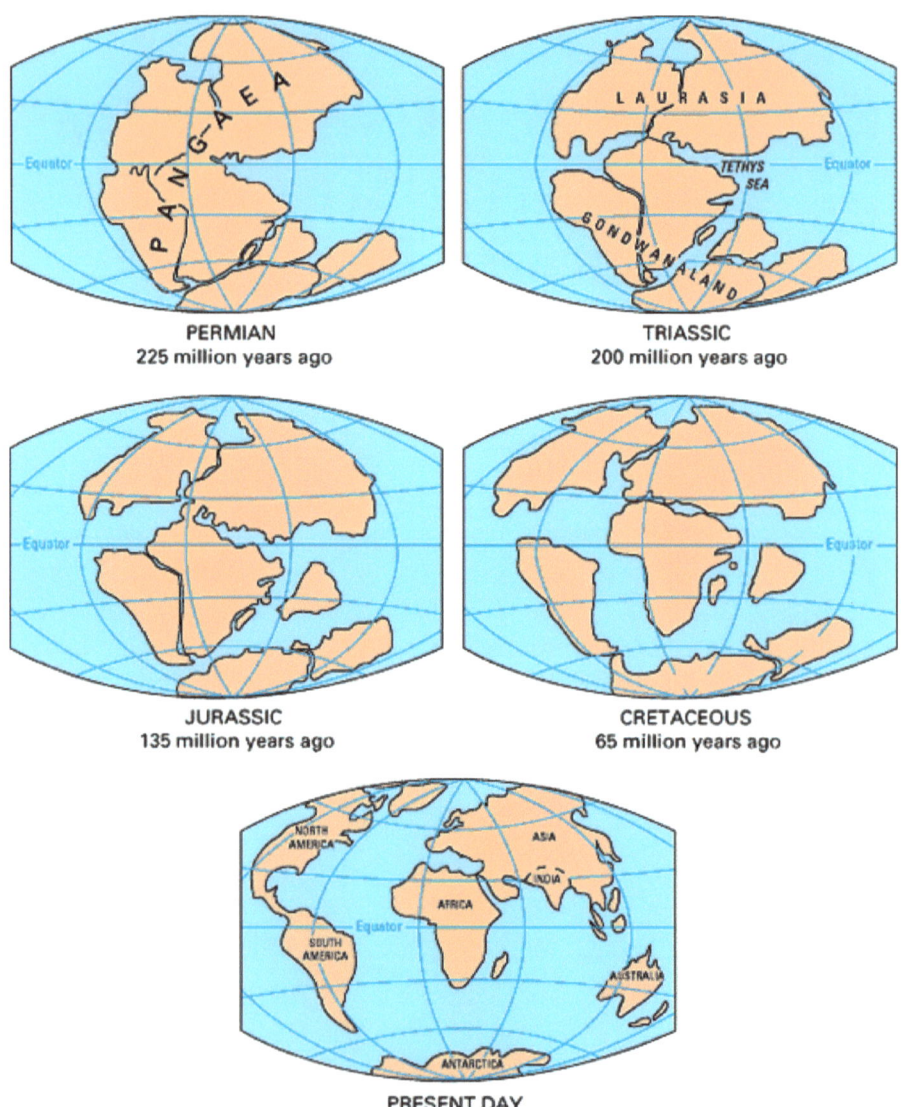

PERMIAN
225 million years ago

TRIASSIC
200 million years ago

JURASSIC
135 million years ago

CRETACEOUS
65 million years ago

PRESENT DAY

And what would that mean for the evolutionary path of each planet? We know Earth used to look completely different only 300 million years ago according to **PANGAEA.**

Is Earth the only planet that is changing, evolving? What would it look like if the other planets in our Solar System are changing as well?

If all the planets have expanding orbits...like our Moon does...If all the planets are changing and evolving like Earth does...What will the planets look like next, after they've been changing for millions of years, and moving further away from the Sun?

The planets are changing ... into each other!

The planet **Mercury...**

...is changing into...
the planet **Venus...**

The planet **Venus...**

is changing into...

the planet...**Earth!**

And the planet **Earth...**

is changing into the planet **Mars!**

NEED MORE PROOF?

Examine the features of each planet and the connections become clear.

SOLAR SYSTEM COMPONENTS + EVOLUTIONARY PATH	FEATURES
SUN	Light/Heat/Solar Flares
MERCURY	Many Impact Craters / Very Small
VENUS	Hottest Planet/Thousands of Volcanoes/Thick, Dense Cloudy Atmosphere
EARTH	Few Volcanoes/Sparse Clouds Abundant Water and Minerals Metals/One Moon
MARS	The "Red" Planet (Reddish-Orange or Rust Colored)/2 Moons
ASTEROID BELT	Millions of Asteroids
JUPITER	"Gas Giant"/Largest Planet Very Gassy Atmosphere/Lots of Moons/Extremely Thin Ring Structure/"Great Red Spot"
SATURN	"Gas Giant"/2nd Largest Planet Very Gassy Atmosphere/Lots of Moons/Extremely Wide Ring Structure
URANUS	"Ice Giant"/Gassy Atmosphere Wide Ring Structure
NEPTUNE	"Ice Giant"/Gassy Atmosphere Faint Ring Structure
PLUTO	Smallest Planet/Icy, Cold Surface Farthest From Sun
WHOLE SOLAR SYSTEM	Sun/Planets And Moons Asteroids/Comets/Cosmic Dust, Debris and Gas

The Mars Curiosity Rover has found evidence of flowing water on Mars—proof that it used to look like Earth.

NASA handout showing water flow channels on Mars.

Mars Curiosity Rover Finds Proof of Flowing Water—A First

NASA reveals unprecedented on-the-ground evidence: telltale pebbles and gravel.

NASA's Mars Science Laboratory (aka Curiosity), as seen via self-portraits of its undercarriage.
Image courtesy MSSS/CaltechINASA

Marc Kaufman
for National Geographic News

Published September 27, 2012
See updated story: Mars Rover Finds Ancient Streambed

NASA's Mars rover Curiosity has made its first major science discovery, and it's one for the ages. Scientists at the Jet Propulsion Laboratory (JPL) in Pasadena, California, announced Thursday that water-fast-running and relatively deep—once coursed over Mars's now bone-dry surface, a finding based on the presence of rounded pebbles and gravel near the rover's landing site in Gale Crater. What's more, the researchers estimate that the water was present for thousands to millions of years.

The finding represents the first proof that surface water once ran on Mars. Planetary scientists have hypothesized that the cut canyons and riverlike beds photographed by Mars satellites had been created by running water, but only now do they have on-the-ground confirmation- and the promise of learning much more about the nature and duration of the water flows.

http://news.nationalgeographic.com/news/20 12/09/12 092 7 -nasa -mars-science-laboratory -... 10/24/2012

The evidence in the newfound streambed led Curiosity lead scientist John Grotzinger—known as a cautious and careful scientist—to conclude that the rover had already found a site that was potentially habitable in the distant past. That doesn't mean life existed there or anywhere else on Mars, he said, but rather that some key physical conditions appear to have allowed for its possible emergence.

The ultimate connection about the **Theory of Interplanetary Evolution** is that…

If **Earth** is evolving into **Mars** ...then **Earthlings** must be evolving into ...**MARTIANS?**

IS THIS THE REASON MARS IS EVERYWHERE IN POP CULTURE?

L'il Wayne produced a video about being a Martian that can be viewed on YouTube.

The Cash Money star explains it's a cool way to say "I'm different."

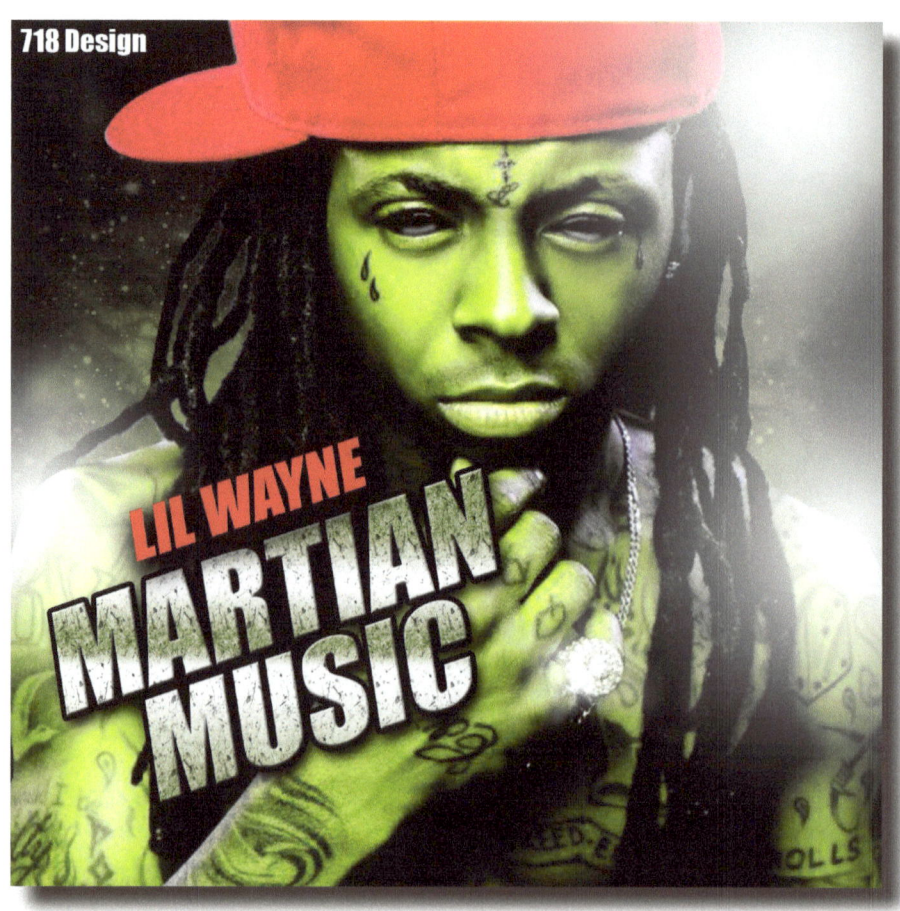

Nikki Giovanni penned a poem about Mars called
We're Going to Mars.

QUILTING THE BLACK-EYED PEA
(We're Going to Mars)

We're going to Mars for the same reason Marco Polo
rocketed to China
>>> for the same reason Columbus trimmed
>>> his sails on a dream of spices
>>> for the very same reason Shakelton
>>> was enchanted with penguins
>>> for the reason we fall in love

It's the only adventure

We're going to Mars because Peary couldn't go to
the North
>>> Pole without Matthew Henson
>>> because Chicago couldn't be a city
>>> without Jean Baptiste Du Sable
>>> because George Washington Carver and
>>> his peanut were the right partners for
>>> Booker T.

It's a life seeking thing

We're going to Mars because whatever is wrong with us
will not
>>> get right with us so we journey forth
>>> carrying the same baggage
>>> but every now and then leaving
>>> one little bitty thing behind:
>>> maybe drop torturing Hunchbacks here;
>>> maybe drop lynching Billy Budd there;
>>> maybe not whipping Uncle Tom to death;
>>> maybe resisting global war.

One day looking for prejudice to slip......one day looking
for hatred to tumble by the wayside.......one day maybe
the whole community will no longer be vested in who
sleeps with whom......maybe one day the Jewish community will be at rest......the Christian community will be

content......the Muslim community will be at peace......and
all the rest of us will get great meals at Holydays and
learn new songs and sing in harmony

We're going to Mars because it gives us a reason to change
If Mars came here   it would be ugly
                nations would ban together to hunt down
                    and kill Martians
                and being the stupid undeserving life
                    forms that we are
                we would also hunt down and kill
                    what would be termed
                Martian Sympathizers
                As if the Fugitive Slave Law wasn't
                    bad enough then
                As if the so-called *War on Terrorism*
                    isn't pitiful Now
When do we learn and what does it take to teach us
    things cannot be:
                        What we want
                        When we want
                        As we want
                Other people have ideas and inputs
                And why won't they leave Rap Brown alone
The future is ours to take

We're going to Mars because we have the hardware to do it
    we have
                        Rockets and fuel and money and stuff
                            and the only
                        Reason NASA is holding back is they
                            don't know
                        If what they send out will be what
                            they get back
So let me slow this down;

Mars is 1 year of travel to get there ....
            plus 1 year of living on Mars ....
            plus 1 year to return to Earth ....
            = 3 years of Earthlings being in a tight space

going to an unknown place with an unsure welcome
awaiting them...tired muscles...unknown and unusual
foods...harsh conditions...and no known landmarks to
keep them human ... only a hope and a prayer that they
will be shadowed beneath a benign hand and there is no
historical precedent for that except this:

The trip to Mars can only be understood through Black
  Americans
I say, the trip to Mars can only be understood through
  Black Americans

The people who were captured and enslaved immediately
recognized the men who chained and whipped them
and herded them into ships so tightly packed there was
no room to turn...no privacy to respect...no tears to fall
without landing on another...were not kind and gentle
and concerned for the state of their souls...no...the men
with whips and chains were understood to be killers...
feared to be cannibals...known to be sexual predators...
The captured knew they were in trouble...in an unknown
place...without communicable abilities with a violent and
capricious species...
But they could look out and still see signs of Home
  they could still smell the sweetness in the air
  they could see the clouds floating above the land
    they loved
But there reached a point where the captured could not
  only not look back
    they had no idea which way "back" might be
    there was nothing in the middle of the deep blue
    water to indicate which way home might be and
    it was that moment...when a decision had to be
      made:
        Do they continue forward with a resolve to
        see this thing through or do they embrace
          the waters and find another world
In the belly of the ship a moan was heard...and someone
picked up the moan...and a song was raised...and that
song would offer comfort...and hope...and tell the story...

When we go to Mars…………..it's the same thing…….
     it's Middle Passage
When the rocket red glares the astronauts will be able to
see themselves pull away from Earth…as the ship goes
deeper they will see a sparkle of blue…and then one day
not only will they not see Earth…they won't know which
way to look… and that is why NASA needs to call Black
America

They need to ask us: How did you calm your fears…How
were you able to decide you were human even when everything said you were not…How did you find the comfort in the face of the improbable to make the world you
came to your world…How was your soul able to look back
and wonder

And we will tell them what to do: To successfully go to
Mars and back you will need a song…take some Billie
Holiday for the sad days and some Charlie Parker for the
happy ones but always keep at least one good Spiritual
for comfort…You will need a slice or two of meatloaf and
if you can manage it some fried chicken in a shoebox
with a nice moist lemon pound cake…a bottle of beer
because no one should go that far without a beer and
maybe a six-pack so that if there is life on Mars you can
share…Popcorn for the celebration when you land while
you wait on your land legs to kick in…and as you climb
down the ladder from your spaceship to the Martian
surface…look to your left…and there you'll see a
smiling community quilting a black-eyed pea…watching
you descend

John Carter is a fictional character, created by Edgar Rice Burroughs, who appears in Burroughs' Barsoom novels. Although he is actually a Virginian from Earth and only a visitor to Mars, he is sometimes known as John Carter of Mars, in reference to the setting in which his major deeds are recorded. His character is enduring, having appeared in various media since his 1912 debut in a magazine serial. The 2012 Disney-made feature film **John Carter** marks the centenary of the character's first appearance.

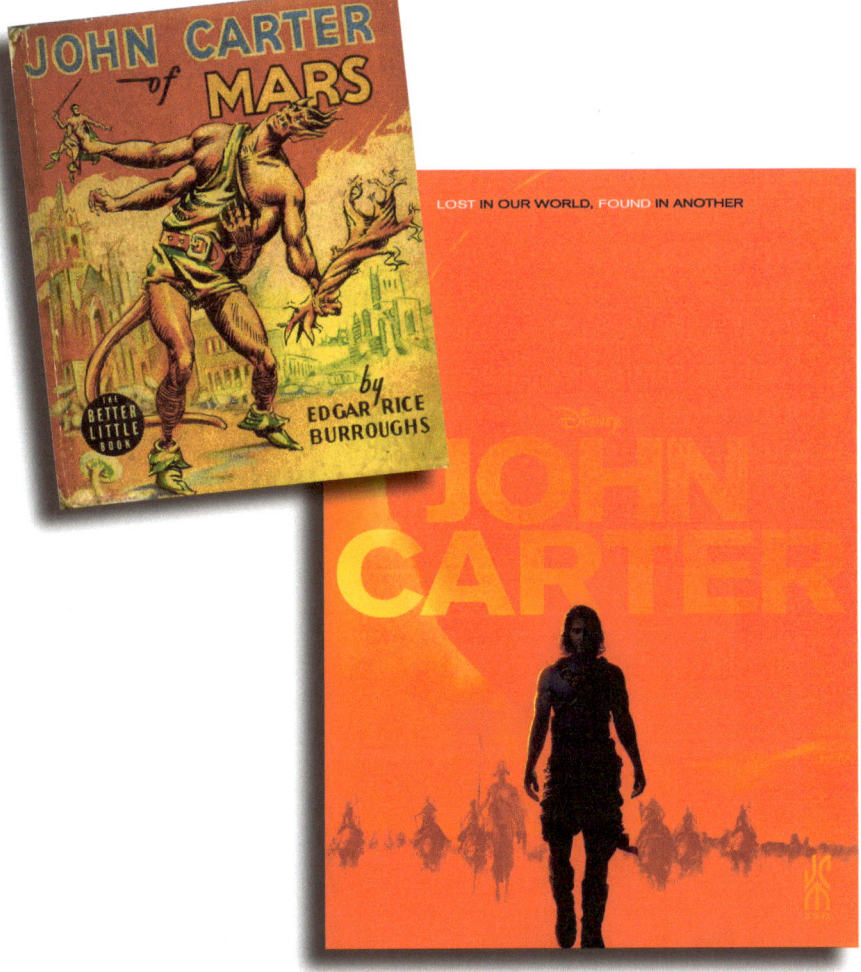

Life on Mars (U.S. TV series)

From Wikipedia, the free encyclopedia

Life on Mars is a science fiction crime drama television series which originally aired on ABC from October 9, 2008 to April 1, 2009. It is an adaptation of the BAFTA-winning original UK series of the same name produced by the BBC. The series was co-produced by Kudos Film & Television, 20th Century Fox Television and ABC Studios.

The series tells the story of New York City police detective Sam Tyler (played by Jason O'Mara), who, after being struck by a car in 2008, regains consciousness in 1973. Fringing between multiple genres, including thriller, science fiction and police procedural, the series remained ambiguous regarding its central plot, with the character himself unsure about his situation. The series also starred Harvey Keitel, Jonathan Murphy, Michael Imperioli, and Gretchen Mol.

Life on Mars garnered critical praise for its premise, acting, and depiction of the 1970s but suffered from a declining viewership after its premiere and a two-month hiatus. ABC announced on March 2, 2009 that it would not be ordering a second season. A DVD set of the complete series was released on September 29, 2009.

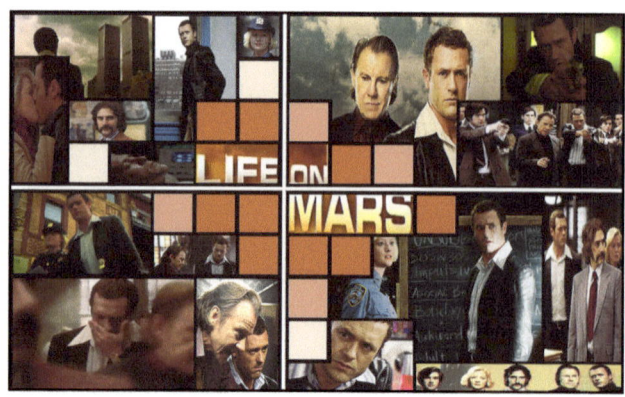

OTHER REFERENCES TO MARTIANS IN FICTION AND POP CULTURE

- *The War of the Worlds* (1898) by H. G. Wells. The Martians are an ancient, advanced race with a tentacled, cephalopod-like appearance, who are invading Earth as their own planet is cooling down.

- *Aelita (1923): Aelita, Queen of Mars*, novel written by Russian writer Alexey Tolstoy. The Martians live in a class-based society; their workers rise up against the ruling class, but the revolution fails.

- Olaf Stapledon's *Last and First Men,* a vast future history published in 1930 and spanning billions of years, includes a long and carefully worked-out account of several Martian invasions of Earth over a period of tens of thousands of years.

- C. S. Lewis wrote, in *Out of the Silent Planet*, about three humans visiting Mars and meeting three different kinds of native intelligent creatures (sorns, (or séroni), hrossa, and pfifltriggi) there, as well as hunting hnakra and meeting the Oyarsa, or eldil in charge of this planet, called Malacandra in the Old Solar language. These Martians are dying out, but resign themselves to their fate.

- Raymond Z. Gallun's **Seeds of the Dusk**, published in 1938, shows the influence of both Wells and Stapledon, but with a special original twist. In this case, the invasion is successful, and it is the Itorloo, distant descendants of mankind, who are exterminated by a plague microbe artificially produced by the invaders.

- In four stories by Eric Frank Russell published in the early 1940s and collected in the classic **Men, Martians and Machines**, humans together with very likable Martians are shipmates who go out together into interstellar space and guard each other's backs while encountering various other aliens.

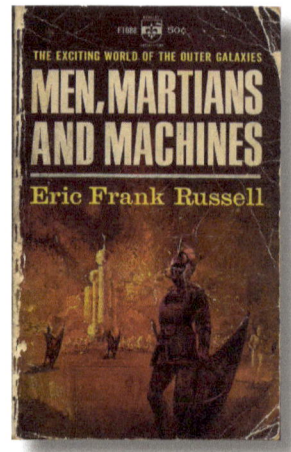

- Ray Bradbury's novel **The Martian Chronicles** depicts Martians as a refined and artistic race of golden-skinned beings who closely resemble humans. They are almost completely wiped out by the diseases brought to Mars by human invaders. At the end of the book the human inhabitants of Mars realize that they are the new Martians.

- Fredric Brown wrote **Martians, Go Home** (1955), a spoof of Wells's Martian invasion concept.

- Larry Niven featured humanoid Martians with a primitive material culture inhabiting an environment of red dust and salpetric acid, most notably in **Protector** (1973), which also includes their genocide.

- Robert A. Heinlein repeatedly used Martians (usually, human beings born and bred on Mars) as characters in his novels and short stories, including *Red Planet* (1949), *Double Star* (1956), and *Stranger in a Strange Land* (1961).

- In D. F. Jones's 1977 novel *Colossus and the Crab,* Martian life predated life on Earth, but faced a process of devolution as conditions on the planet worsened.

  - **Monument** commemorating where the Martians "landed" in West Windsor, New Jersey

  - The October 30, 1938, radio broadcast of *The War of the Worlds* was the cause of much confusion when it was aired, with some people believing an actual Martian invasion was taking place.

- Looney Tunes included the cartoon character **Marvin the Martian** (1948), a comic foil to Warner Bros. mainstays Bugs Bunny and Daffy Duck in several animated shorts. He attempts to blow up the earth, as it "obscures his view of Venus". Later, he appears as the Martian Commander in **Duck Dodgers in the 24-1/2th century.**

- *Red Planet Mars* (1952) – Scientist Peter Graves contacts Martians by radio; they respond by preaching Christianity, and thus communism is defeated.

- The Twilight Zone – **"Will the Real Martian Please Stand Up?"** has Martians attempting to colonize Earth. A humanoid Martian appears with three arms. However, the colonization is prevented by Venusians.

- *Invaders from Mars* (1953) – A film, remade in 1986.

- *Quatermass and the Pit* (1958–1959) – A British television serial in which a crashed spacecraft is discovered in London, which reveals that humanity on Earth is the result of experiments by a Martian civilisation, now long dead. It was remade as a film in 1967.

- **My Favorite Martian** (1963–1966) – A television comedy series and film.

- **Doctor Who** includes a race native to the planet Mars known as the Ice Warriors, whose planet is dying out. The show also features a Martian virus based within the planet's water.

- **Captain Scarlet and the Mysterons** (1967–1968) – The Martians at war with Earth are the Mysterons — an invisible race of superbeings hell-bent on revenge after an unprovoked attack on their Martian city by Captain Black, a Spectrum agent investigating strange alien signals.

- **Spaced Invaders** (1990) – A sci-fi comedy in which dim-witted Martians attempt to invade a small Illinois town during a re-broadcast of Orson Welles's 1938 War of the Worlds.

● **Total Recall** (1990)– A science fiction action film starring Arnold Schwarzenegger, where the plot concerns an apparently unsophisticated construction worker who turns out to be a freedom fighter from Mars and has been relocated to Earth. He later learns of an alien artifact proving life had previously existed on Mars.

● Philip K. Dick used the planet as a setting for many of his novels. In **Martian Time-Slip**, a human colony is trying to cope with arduous environmental conditions and there is also an aboriginal race of "Bleekmen" who are treated with casual racism. In **The Three Stigmata of Palmer Eldritch**, there are no indigenous inhabitants; so given the arduous ecological context, human colonists are dependent on drugs like "Can-D" and "Chew-Z."

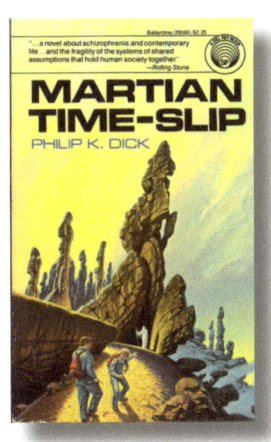

● **Mars Attacks!** (1996) – A satirical film directed by Tim Burton, based on the equally satirical Topps trading card series Mars Attacks (1962).

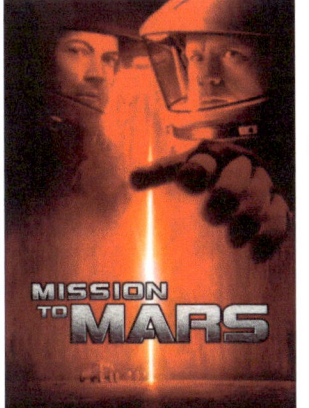

● **Mission to Mars** (2000) – Martian(s) are shown as tall, feminine and very peaceful humanoids who abandoned their planet after a large meteor struck.

- **Ghosts of Mars** (2001) – Humans battle Martians for life on Mars.

- In the Invader Zim episode **"Battle Of The Planets"** (2001) Zim discovers that the Martian race became extinct after transforming the planet Mars into a giant space ship.

- **Mars Needs Moms** (2011) The male Martians have been banished from the planet by the females.

- In the DC Comics universe, the **Martian Manhunter** (J'onn J'onzz) (1955) is a superhero and a member of the Justice League, believed to be the last of the peaceful Green Martians. Other DC creations include **Miss Martian** and the  **White Martians**, the enemy of the Green Martians.

- In the **Adventures of Superman** story **"Black Magic on Mars"** from issue #62 (January 1950), Martians led by the dictator Martler, an admirer of Hitler, appear.

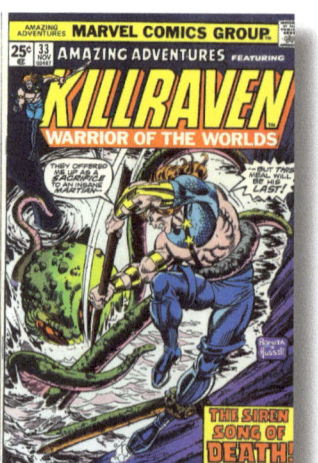

- In the future world of Marvel Comics' **Killraven** (1973), the Martian Masters who orchestrated the invasion in The War of the Worlds returned to Earth a century later and conquered it; they were overthrown by rebels led by the psychic human Jonathan Raven, alias Killraven.

- In the *Metal Slug* series, the Mars People are common enemies and plot devices very similar to the ones described by H.G. Wells.

- In the turned-based tactics game **X-COM: UFO Defense**, the alien invaders use Mars as a base of operations in which to launch UFO attacks on Earth.

- In the video game **Stalin vs. Martians** (2009), one plays the leader of the Soviet Union taking charge of defending the Earth from invading Martians.

- **Doom 3** features Martians as an ancient extinct race of people.

- The 1962 trading card series *Mars Attacks* (no exclamation point, unlike the 1996 film based on it) depicts an invasion of Earth by hideous, skeletal Martians.

THE INVASION BEGINS

MARS ATTACKS

SPACE ADVENTURE BUBBLE GUM 5¢

- The Misfits have various songs related to Martians, e.g. **"Teenagers from Mars"** and **"I Turned Into a Martian."**

- Rebecca Bloomer's novel *Unearthed* (2011), the first in a series, depicts a futuristic human colony on Mars. A distinction is made between those born on Earth who immigrated to Mars and the local Martians who were born there and have never known any other home.

LIFECYCLE POEM

Gerald Nordé, Jr .

I ain't even here yet
But I'm going to rock this life you can bet

Even as a kid I am impressive
My genius is so massive it can't be measured

I make it past ten, trying to get to twenty
Don't have no money but trying to get plenty

Trying to get fat pockets like electricity from sockets
IRS audits will be so shocked they'll want to stop it

I'm past thirty, I've made a mess
But still in pursuit of boundless success

Tough times for the people ahead, but I won't dread
No weapon shall prosper like Isaiah once said

My back has a heavy load
But I can bear it; the Alpha and Omega has paved the road

I'm going to look back and be celebrated
All the ways that I made it, everything was A-rated

Wishing I could do it again because everything was a win
How will I be judged though when we remember the sin

My worries are with a fury as another realm allures me
But I'll keep putting in work so I'll make the afterlife weary

Death should fear me when it gets near me
It might not hold me for long

Everyone that's already gone can't wait for me to arrive
Because they love the way I'll do it all again...LIVE!!

www.ingramcontent.com/pod-product-compliance
Lightning Source LLC
Chambersburg PA
CBHW041143180526
45159CB00002BB/710